To my dad, Bill, who loved gardening and grew the most beautiful flowers from seed. Love you, Dad. x

DEBRA WELLINGTON

CANDY JAR BOOKS

This is Sid.

    Sid is a sunflower seed. He's small and grey, but wants to be tall and bright like the flowers in his garden.

"Why can't I be tall and colourful like you?" said Sid to the flowers.

"You need to be patient, Sid. You'll have to wait to grow," said the flowers.

"That will take forever. I want to be tall NOW!" said Sid.

"Snuggle down into the soil and wait for the sun and rain. Then you will start to grow," said the flowers.

So Sid did just that.

   He snuggled down deep in the soil and waited.
Soon the sun came out and warmed up the soil.

   "This is lovely and cosy," said Sid.

An earthworm wriggled by on its way through the soil.

"Hey, what are you doing?" asked Sid.

"I'm making the soil better for you to grow. I eat up all the dead plants and keep the soil fresh and airy," replied the earthworm.

"That sounds great!" said Sid.

"I also poop into the soil, which makes the soil good for you to grow in," said the earthworm.

Then it rained and rained, deep into the soil.

    It wasn't long before Sid noticed that he was beginning to sprout roots. These were like drinking straws for plants.

    Sid used his roots to suck up the water and food from the soil, so that he would grow quicker and stronger.

The roots kept Sid anchored deep into the soil.

After a while, a simple stem grew, poking up above the soil.

Then two leaves grew from the stem.

Sid was now a seedling. He was growing into a sunflower plant!

Sid could now feel the warmth of the sun.

"What's this round thing?" asked Sid, as the rain continued to fall onto his leaves.

"It's your flower bud. You're starting to grow your sunflower," said the flowers.

"Yay!" shouted Sid.

"Now listen, Sid," said the flowers. "You need lots of sunshine to grow tall and strong, so keep facing the sun as you grow."

Days and weeks went by and Sid kept on growing.

His flower bud grew bigger and bigger, like a big round button. It started to open up slowly, and showed all its bright yellow petals.

"Look at me! I'm a sunflower, and I'm even taller than you," said Sid to the flowers.

"You are tall and colourful just like you wanted, Sid," said the flowers.

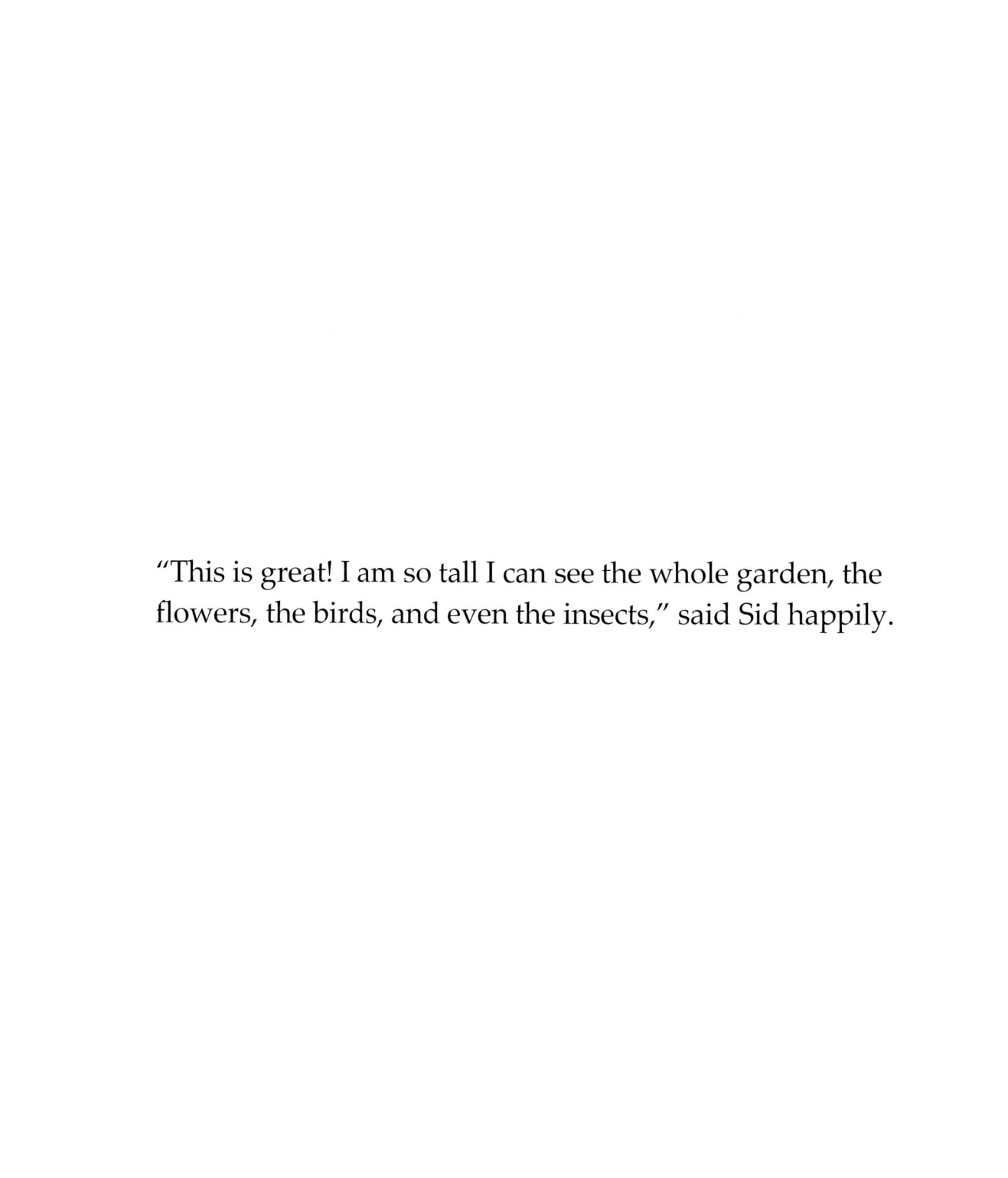

"This is great! I am so tall I can see the whole garden, the flowers, the birds, and even the insects," said Sid happily.

Just then a bee flew and landed in the middle of Sid's flower.

"Hey, what are you doing?" said Sid to the bee.

"Yummy! I love to eat the sweet nectar in your flower, and I feed the pollen to my children" the bee replied.

As the bee flew away, Sid noticed that some of the pollen had stuck to the bee's furry body and legs.

The pollen would be passed onto the next flower the bee landed on. This is called pollination and is very important as it means that plants can make new seeds to grow into new plants.

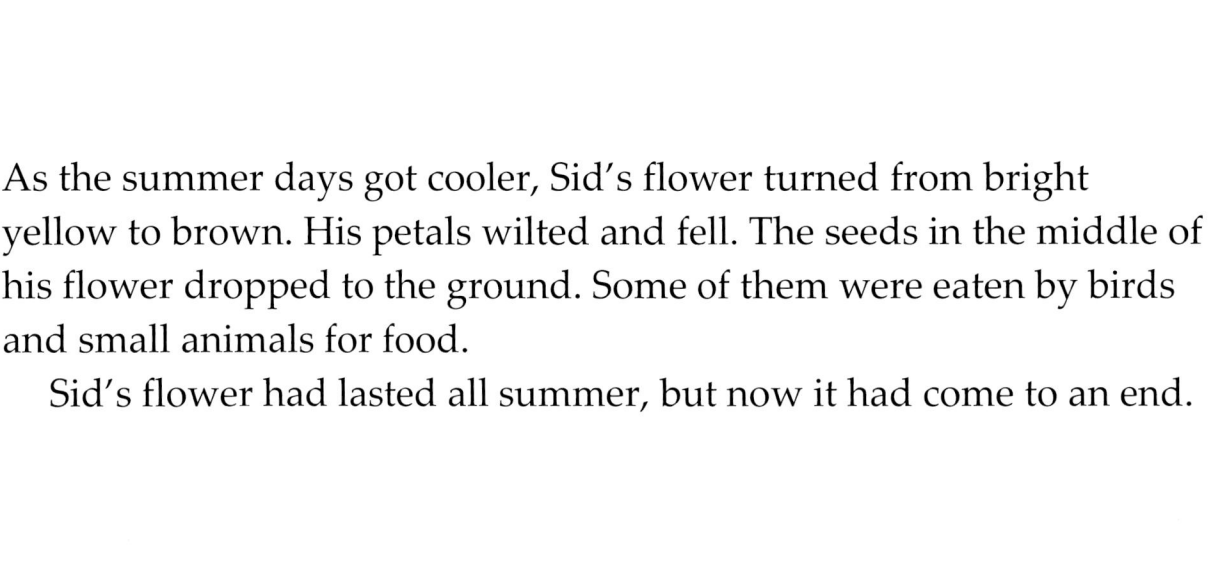

As the summer days got cooler, Sid's flower turned from bright yellow to brown. His petals wilted and fell. The seeds in the middle of his flower dropped to the ground. Some of them were eaten by birds and small animals for food.

Sid's flower had lasted all summer, but now it had come to an end.

Some of the seeds fell deeper into the soil, snuggling down just like Sid had done.

As time passed, the rain came, the sun shone, and the seeds lay quiet under the soil.

But look…

Now they have grown roots, leaves and are little seedlings.

They will grow into tall and bright sunflowers, just like Sid.